Common Sense Guide to Fire Safety and Management

An essential and short guide for those who need to know more about fire safety and management without wanting to spend hours reading dozens of different documents. Whether it's for use alongside a training course or simply to brush up on your knowledge, it's perfect for equipping you with the principles of fire safety.

Friendly and accessible, this *Common Sense Guide* covers all the main aspects of fire safety and management in manageable chapters to provide you with the knowledge and understanding you need to look after yourself and others.

- Suitable for those with little understanding of fire safety and management
- Includes questions at the end of each module to consolidate your fire safety knowledge
- Certificate offered to those who complete the exam at the end of the book and return to be marked externally.

Subash Ludhra is a past president of The Chartered Institution of Occupational Safety and Health (IOSH) and considered to be an expert in the field of Risk Management. Having qualified as an Occupational Hygienist, Subash Ludhra now manages Anntara Management Ltd; an international risk management and loss control consultancy business that operates in the UK and overseas.

COMMON SENSE GUIDES TO HEALTH AND SAFETY

Common Sense Guide to Health and Safety at Work
978-0-415-83544-2 February 2014

Common Sense Guide to Fire Safety and Management
978-0-415-83542-8 February 2014

Common Sense Guide to Environmental Management
978-0-415-83541-1 February 2014

Common Sense Guide to International Health and Safety
978-0-415-83540-4 September 2014

Common Sense Guide to Health and Safety in Construction
978-0-415-83545-9 October 2014

Common Sense Guide to Health and Safety for the Medical Professional
978-0-415-83546-6 November 2014

Common Sense Guide to Fire Safety and Management

Subash Ludhra

Routledge
Taylor & Francis Group

LONDON AND NEW YORK

First published 2014
by Routledge
2 Park Square, Milton Park, Abingdon, Oxon, OX14 4RN

and by Routledge
711 Third Avenue, New York, NY 10017

*Routledge is an imprint of the Taylor & Francis Group,
an informa business*

British Library Cataloguing in Publication Data
A catalogue record for this book is available from the British Library

Library of Congress Cataloging-in-Publication Data
Ludhra, Subash.
    Common sense guide to fire safety and management /
    Subash Ludhra.
        pages cm — (Common sense guides to health and safety)
    Includes bibliographical references and index.
    1. Office buildings—Fires and fire prevention. I. Title.
    TH9445.O4L83 2014
    658.4'77—dc23   2013038670

ISBN13: 978-0-415-83542-8 (pbk)
ISBN13: 978-1-315-85876-0 (ebk)

Typeset in Sabon
by Keystroke, Station Road, Codsall, Wolverhampton
Printed and bound in Great Britain by Ashford Colour Press Ltd

Contents

Contents

Foreword

As the Chief Executive Officer of a large group of Companies, I can appreciate the vital importance of Fire Safety Management, to control risks and to protect our people and operational assets from the effects of fire.

Fire can be devastating; for the people caught up in an event and also for the company or operator of the premises. Quite apart from the obvious risk to precious life, the disruption and damage that can be done can be difficult to calculate.

It is a fact that many people and companies never recover from a major fire but even more important than that is the ongoing effect that it has on our people, in the short, medium and long term.

Therefore I believe that the case for effective management of Fire Safety is undeniable and a significantly important part of that is the need for ongoing high-quality information, instruction, training and understanding of the issues.

This *Common Sense Guide to Fire Safety and Management* is an excellent reference for those individuals with little or no prior knowledge of Fire Safety and Fire Safety Management who need to know more about them. This guide is offered to support basic training without being too specific.

Any source of information that helps to improve our knowledge of fire safety in an easy to understand format is to be commended.

It is an honour and a privilege to be asked by the author to write this foreword and I wish you well as you strive each and every day to improve fire safety and reduce fire risk.

John Shropshire – CEO
The Shropshire Group

Welcome

It is often said that we all know what is right and what is wrong, what we should and should not do. Surely it's just "common sense". Unfortunately common sense is not as common as we would like to believe. The definition of common sense is "the ability to behave in a sensible way and make practical decisions". Most individuals and employers will at some time do things that are not sensible or practical and some will do this regularly.

This guide has been developed to help improve your knowledge of Fire Safety and the management of fire risks within the workplace in a light-hearted way. It is designed to further heighten the common-sense element of your existing knowledge.

Although the guide refers to UK legislation/best practice, the principles are applicable internationally. However, you should answer all questions within the guide, based on the UK legislation.

By taking the time to improve your knowledge and learn more about fire and fire safety in the workplace you can:

- **avoid having a fire in the workplace**
- **know what to do in the event of a fire outbreak**
- **help to prevent fires spreading**
- **carry out a simple fire-risk assessment**
- **make your workplace or home safer.**

Every year, approximately 30 people are killed (directly or indirectly) while at work through fires and thousands more are injured. The cost to industry and society as a whole is huge.

Most importantly, if you are injured as a result of a fire or suffer ill health you and your family are likely to suffer.

DID YOU KNOW?

- Each year around 30 people are killed as a result of work-related fires.
- Annually over 2,500 people are injured as a result of fires.
- The UK fire and rescue services attend over 35,000 fires at work every year.
- Fire and explosion incidents account for approximately 2% of the major injuries reported to the enforcement authorities.
- Fires start and spread much quicker than most people expect.

By working through this accessible guide you will learn more about fire safety in the workplace and as a result you will be better placed to recognise the hazards and dangers associated with fire in any workplace or even your own home 24 hours a day.

Fire

How this guide works

AIMS

The aim of this guide is to provide you with a basic understanding of:

- the principles of fire
- how fires start and spread
- fire risk assessment
- the principles of fire prevention
- fire evacuation
- fire alarms, detection and fire fighting.

OBJECTIVES

By the time you finish this guide, you will be able to:

- explain the basic principles of fire
- define hazard and risk
- help reduce the risk of a fire starting or spreading within your workplace
- identify sources of ignition/heat within the workplace
- identify sources of fuel within your workplace
- identify sources of oxygen within your workplace
- know what to do in the event of a fire
- explain how fires spread
- identify the main dangers associated with fire
- explain how fires can be extinguished
- explain how fire can be detected
- understand the fire-risk-assessment process
- explain the role of the fire marshal/warden.

How to complete the guide

Before you start to complete this guide please read the notes below in order to ensure that you get the most out of your training.

WHAT DO YOU NEED TO COMPLETE THE GUIDE?

You will need:

- a quiet cosy environment that allows you to relax and make notes;
- a pen and paper to make notes;
- a desire to learn and improve your knowledge of fire and fire safety.

GUIDANCE ON LEARNING

This guide has been produced to help you learn more about fire and fire safety in your workplace and to reduce the likelihood of a fire starting and spreading or causing injury or ill health. The booklet allows you to complete your studies at your own pace with the support of your manager or supervisor. However, we recommend that you complete the guide within four weeks.

Throughout the guide there are simple questions designed to help you test your subject knowledge and learn.

If you cannot answer the questions please read the relevant topic again to refresh your memory. If you are still in doubt please speak to your line manager who will be able to assist you.

When you have completed the course booklet and the exercises, you may wish to complete the examination

(20-question multiple-choice exam paper on pages 87–93). On achieving the required pass mark (75%) you will receive a certificate to confirm that you have completed the guide and passed the associated examination.

HOW TO USE THE GUIDE

The guide is divided into modules. We recommend that you complete one module at a time in full, starting with module one, progressing sequentially through to the last module. You do not have to complete the guide in one sitting. A lot of information is provided and you may learn more effectively by tackling the modules in bite-sized chunks.

The modules are designed to take you through a specific learning pattern to help you learn. Each module contains questions to make you think about your own job and workplace. There are also questions at the end of each module to test your knowledge and understanding of it. The answers can be found on pages 77–82 of the guide. There may be times when you feel you need help and support in completing the guide. Should this be the case please speak to your line manager.

Employers and employees must work together

WHEN DO I GET MY CERTIFICATE?

Once you have completed the guide and the examination paper, the paper will be marked and on achieving the required pass mark a certificate pdf will be emailed to you as soon as possible.

Remember the certificate only confirms that you have completed the guide and passed the associated exam. The real benefit to you will come from your improved knowledge and ability to identify fire hazards and reduce the risk of having a fire at work or in the home. **Good Luck!**

Winning trophy

MODULE ONE

Fire

MAJOR FASHION RETAILER FINED £400,000 FOR BREACHES OF THE REGULATORY REFORM FIRE SAFETY ORDER FOLLOWING A MAJOR FIRE AT THEIR OXFORD STREET STORE.

Welcome to module one. In this module you will learn more about fire, what causes it and the hazards it presents within your workplace.

All materials within your workplace have the potential to burn (under the right conditions) and it is everybody's responsibility to ensure that all fire risks are minimised or avoided by complying with fire procedures.

WHAT IS FIRE?

Fire is a chemical reaction called combustion, resulting in the release of heat, light and smoke. The following three elements are essential for the starting of a fire:

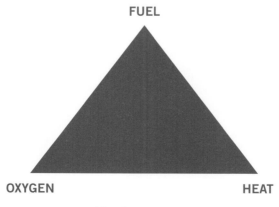

The fire triangle

FUEL (Everything in the workplace is potentially flammable.)

OXYGEN (Air in the workplace contains 21% oxygen which we also need to survive.)

HEAT/IGNITION (This could be direct from a flame or indirect from a heater or even sunlight.)

The combination of all three elements in an uncontrolled manner presents a fire risk.

SOURCES OF HEAT/IGNITION

There are numerous sources of heat within any workplace. Some more obvious than others. Examples include:

- **external sparking** of machinery (grinding wheels, welding);
- **naked flames** (smoking materials, cooking appliances, heating appliances);
- **internal sparking** (electrical equipment / switches);
- **hot surfaces** (lighting, cooking, heating appliances, poorly ventilated equipment);
- **static electricity** creating high voltage sparks (pouring flammable liquids).

Q

List four sources of heat or ignition within your work area.

SOURCES OF FUEL

The workplace is full of sources of fuel. These will be one of three forms:

- **solids** (wood, paper, cardboards, packaging, plastics, rubbers, foam, textiles, waste materials);
- **liquids** (varnishes, paints, adhesives, petrol, white spirit, paraffin, methylated spirits and toluene) – remember most flammable liquids will give off vapours which are heavier than air so will accumulate at the lowest level;
- **gases** (liquefied petroleum gases, acetylene, hydrogen).

Fuel can

SOURCES OF OXYGEN

As well as the air which contains oxygen (21%) additional sources of oxygen in the workplace may include oxygen cylinders and chemicals which liberate oxygen. Ventilation systems provide a constant supply of air in the event of a fire unless they are closed.

By removing or controlling any one of the elements, the risk of a fire starting or spreading is reduced or eliminated.

Q

What flammable materials are stored/used in your place of work?
List at least five examples

_____

_____

_____

_____

_____

_____

_____

COMMON CAUSES OF FIRE IN THE UK

The most common causes of fire in the United Kingdom are:

Arson

Thoughtless individuals cause most fires deliberately in the UK.

Always report anybody seen acting suspiciously and do not make it easy for the arsonist by carelessly leaving combustible materials out to burn – it could be your life at risk.

Electrical equipment

Faulty electrical equipment can provide a source of heat capable of starting a fire. Always report any equipment you suspect of being faulty or damaged to your line manager.

Potential overloading of sockets

Faulty electrical wiring

All equipment should be visually checked for frayed or damaged cabling prior to use. Always report any defects seen to your line manager and never overload plug sockets.

Mis-use of smoking materials

Discarded cigarette ends can provide a source of heat for potential fire.

Always obey your company's smoking rules, smoke only in designated areas (where provided) and always ensure cigarette ends and matches are disposed of safely.

Naked flame tools

Gas burners, torches or welding equipment used or left carelessly can cause fires. Only staff trained in the correct safe working practices must be allowed to use naked-flame tools. If you use this type of equipment, take extra care.

Gas

Leaking gas appliances (cylinders, bottles or mains supply) or gases built up in sewer or drainage systems can cause fires or explosions.

Always report any suspected gas leaks in and around your workplace.

Hot surfaces

Hot surfaces within the workplace can over time generate enough heat to start a fire.

Explosions

Highly flammable materials must be stored in a controlled environment at all times, away from sources of heat.

Q

Where is the designated smoking area within your workplace?

_____

_____

_____

DANGERS ASSOCIATED WITH FIRE

Once started fires can be devastating as they can cause:

Burns

The heat and flames from a fire can easily burn human skin, tissue and hair, causing pain, suffering and often permanent scarring.

Smoke inhalation

Depending on the material being burnt, fires generate smoke. When breathed in, this smoke can cause breathing problems; smoke can also impair visibility when trying to exit a building.

Toxic fumes

Some burning materials give off toxic fumes, which can damage the lungs, cause difficulty in breathing or even death.

Structural damage to buildings

The heat generated by fires can be so intense that even building materials are weakened or destroyed, causing structural damage to the workplace.

Never underestimate the power of fires.

Environmental damage

As well as the more obvious effects of fire, it can also cause extensive damage to the environment (gases into the air, water runoff into land or watercourses), disruption to transportation systems as a result of proximity to roads, rail or even airports and can ultimately cause businesses to close down.

Buncefield fire

HOW DO FIRES SPREAD?

Once started, fires spread in a number of ways or combinations. The common methods of fire spreading are as follows:

Convection

As the hot air above the fire rises the intense heat is transferred from one point to another and can cause the fire to spread.

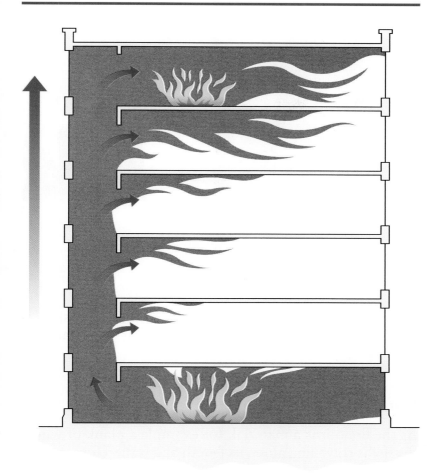

Convection – hot air rising

Radiation

Heat energy is radiated from its source outwards. This can ignite other flammable materials present (remember most materials will burn under the right conditions).

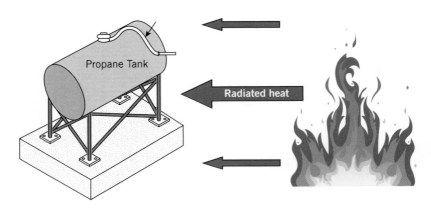

Radiated heat

Conduction

Intense heat can be transferred along a conductor, such as a metal bar, potentially igniting flammable materials further along the bar.

Burning materials

As the fire burns a material, for example a plank of wood, the fire is travelling along its length and is therefore spreading.

EXERCISE 1 – FIRE

Self-assessment questions

To assess your level of understanding please complete the following exercise.

1. What three elements are needed to start a fire?

    a) ....

    b) ....

    c) ....

2. By removing or controlling one of the elements what happens to the fire?

    ....

    ....

    ....

3. List three common causes of fire:

    a) ....

    b) ....

    c) ....

4. List three examples of sources of heat/ignition:

    a) ....

    b) ....

    c) ....

5. What are two of the dangers associated with fires?

    a) ....

    b) ....

6. List three examples of sources of fuel:

    a) ....

    b) ....

    c) ....

7. Give two examples of how fires can spread:

    a) ....

    b) ....

You will find the answers on pages 79–80.

Fire detection and protection

SUPERMARKET GIANT FINED OVER £200,000 FOR SERIOUS FIRE SAFETY BREACHES

In this module you will learn more about the action necessary for successful fire prevention, control and evacuation (should it be necessary).

STRUCTURAL DESIGN PRECAUTIONS

New or refurbished buildings can be designed and built with state of the art building materials to minimise the risk of fires starting or spreading. Existing or older buildings may be harder to protect.

Elements to consider include:

- ensuring occupants can escape from the building safely;
- minimising the spread of fire or smoke;
- maintaining the building's structural integrity for as long as possible.

FIRE DETECTORS AND ALARM SYSTEMS

In any building, being able to detect the presence of a fire and then alerting others as quickly as possible is imperative to save lives and minimise damage.

FIRE DETECTORS

Smoke or heat detectors can be used as early warning systems to detect fires in their early stage of development, allowing time for a prompt response.

Always ensure that you have smoke alarms in your home and check them regularly; they could help save your life.

Fire detector

Smoke detectors

These typically consist of ionisation detectors, where the flow of ions is disrupted by smoke particles, or optical detectors where smoke obscures or scatters a light beam.

Heat detectors

These typically consist of optical detectors sensitive to temperature changes or infra-red/ultra-violet detectors.

Q

Are there any smoke or heat detectors in your area of work? If so, where are they?

_____

_____

_____

_____

_____

_____

_____

_____

_____

_____

ALARM SYSTEMS

Alarm systems can be very simple in low-risk environments or they can be very sophisticated and elaborate for larger complex buildings.

Manual systems

Systems such as word of mouth and/or the use of hand bells, whistles, hand strikers, rotary gongs and hand-operated sirens can be used in small work areas with few people present. They require little or no maintenance but are completely reliant on human input.

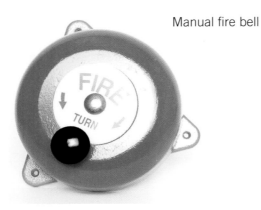
Manual fire bell

Automatic systems

These are linked to fire detectors and can activate alarms, sirens, door closers, extraction fans, communication systems and so on. These systems work automatically and must be regularly maintained and must never be tampered with.

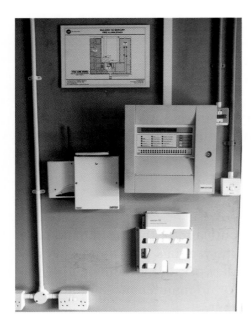
Fire panel

FIRE-FIGHTING EQUIPMENT

Sprinkler systems, fire hoses, fire extinguishers and other fire-fighting equipment can be used to quickly extinguish or help control fires once detected. The use of fire-fighting equipment must however be restricted to competent trained staff only. Never attempt to tackle a fire if you have not been correctly trained to do so. Your employer is responsible for ensuring that all equipment is regularly inspected and you can help by keeping it clear at all times and reporting obstructions or defects.

Fire extinguishers

Different fire extinguishers are designed for use on different types of fire. This is why only trained people should attempt to put out fires.

Using the wrong extinguisher could make the situation worse and put you or your colleagues in danger.

Fire extinguishers work in one of four ways.

- **Cooling** – this reduces the ignition temperature by taking the heat out of the fire.
- **Smothering** – this reduces the amount of oxygen available.
- **Starving** – this limits the amount of fuel available.
- **Chemical reaction** – this interrupts the chain of combustion

Note: All modern cylinders are now red with 5% of the canister showing the colour relating to its contents. Your employer should provide you with appropriate fire training so that you:

- know your site fire/evacuation procedures – if in doubt ask your manager;

- know where your nearest fire exit points are;
- know where your nearest fire alarm call points are located;
- know where your fire extinguishers are located;
- can raise the fire alarm by striking the nearest call point if you see a fire;
- leave the building by the nearest safe exit in an orderly manner (never waste time attempting to collect personal belongings);
- assemble at your appointed fire assembly point;
- do not re-enter the building until an authorised person tells you it is safe to do so.

Q

Do you know where your nearest fire extinguishers are located? Write the locations of the nearest three here.

Material burning	Correct extinguisher to use
Wood/paper/textiles	Water
Flammable liquids	Foam/carbon dioxide
Electrical equipment	Carbon dioxide/dry powder
Cooking oils and fats	Wet chemical/fire blanket

Carbon dioxide and foam fire extinguishers   Foam fire extinguisher

Carbon dioxide and foam fire extinguishers

Fires are usually classified in the following way depending on the material burning:

Class	Fuel
CLASS A	The burning of wood, paper or solid materials
CLASS B	The burning of any flammable liquids such as petrol, oil
CLASS C	Fires involving gases or liquefied gases
CLASS D	Fires involving metals such as aluminium or magnesium
CLASS F	Fires involving cooking oils and fats

Note

Electricity is not a fire; however, electricity can provide the initial heat/spark for ignition. In these situations always try to isolate the electrical source from the mains supply.

Never block fire extinguishers

Sprinkler systems

Sprinkler systems are often used in buildings where it is important to protect the building, contents or occupants – for example hotels and storage warehouses. The sprinkler system heads once activated will release water under pressure to contain the fire until further assistance arrives. Accidental releases can cause significant damage to stored items. These systems must be maintained and regularly serviced.

Fire sprinkler head

Gas systems

A number of gas systems are available to extinguish fires. They are usually used in rooms housing computer equipment or archive materials that need to be protected. They are deemed to be clean forms of extinguishing media as they do not generally damage the stored items. These systems when triggered will normally activate an alarm, giving occupants time to evacuate the room prior to releasing the gas (which can be harmful to human health). If these types of system are used within your workplace ensure you understand the way they work and what action you need to take if the alarm is activated.

NEVER TAKE RISKS WITH FIRE!

They usually grow and spread much quicker than you think they will.

Do not attempt to extinguish a fire unless you are trained to do so, and then do so only if it is small enough to be contained – only attempt to extinguish a fire if it is controllable or is blocking your only means of escape.

- Never block fire escape routes.
- Never block fire escape exit doors.
- Never block fire extinguishers or alarm call points.
- Never prop open internal fire doors.
- If you see a blocked exit route, clear it and report it.
- Always report any damaged or spent fire extinguishers.
- Never smoke in No Smoking areas.

FIRE MARSHALS/WARDENS

Within your workplace your employer may have nominated fire evacuation marshals or wardens. They have additional duties, which may include:

- acquainting themselves with all members of staff within their work area;
- ensuring staff know who they are and what their role is within an emergency situation;
- being familiar with the site evacuation procedures;
- being familiar with the siting and use of fire extinguishers;
- ensuring that all internal fire doors are kept closed;
- carrying out regular inspections of fire-fighting equipment and ensuring all equipment is accessible and exits are kept clear;

- reporting all defective and missing equipment to their supervisor;
- in the event of an evacuation ensuring that their area is evacuated;
- ensuring that the fire brigade has been called;
- carrying out a roll call and reporting as necessary to the nominated fire controller.

If you are a fire marshals or warden, please ensure that you carry out your role as defined; however, never take unnecessary risks like entering a building to carry out your duties if you are already outside.

Q

When was the last time your employer carried out a fire drill? Did everybody leave the building in a timely manner?

EMERGENCY ROUTES AND EXITS

Employers must ensure that all designated exit routes and exits are kept clear at all times. They must be wide enough to allow unrestricted access and the walls should be able to provide adequate protection from fire. You can assist by helping to keep these routes clear and never blocking them.

Exit route

COMPARTMENTALISATION

As one of the methods of a fire spreading is burning materials, compartmentalisation helps to slow down the process of a fire spreading from one area to the next by creating barriers that a fire has to burn through to spread.

TRAINING

Employers are required to identify specific and general training needs for employees and others with specific fire / explosion-related duties. All training provided must be suitable for the audience and refresher training should be provided as necessary.

If you have not been trained or you feel you are in need of refresher training you must inform your employer.

EMERGENCY LIGHTING

Although emergency lighting is not a legal requirement, employers must as part of their fire-risk assessment assess

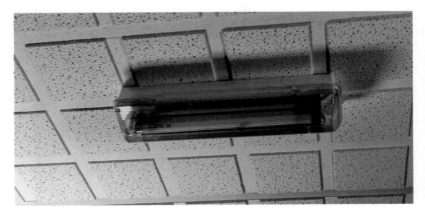

Emergency lighting unit

the need for emergency lighting both inside and outside the workplace. If from the assessment it is deemed necessary then a means of illuminating the exit routes and exits must be installed/made available. There are several ways of achieving this, including traditional emergency lighting units as well as hand-held torches and rods which emit light when snapped. Any emergency lighting provided must be tested and maintained by the employer.

SIGNAGE

From their risk assessment, employers must ensure that adequate and appropriate signage is in place where deemed necessary to assist in the evacuation process. This could include:

- signs above exit doors;
- directional arrows to show directional route;
- signs showing location of fire equipment;
- signs labelling flammable/explosive materials;
- signs along exit routes;
- signs on exit doors.

All signage fitted must be compliant with relevant British Standards and any signage present must be kept clean and be visible at all times.

Know what you need to do

Follow designated exit routes

Fire exit sign

Q

Where is the nearest fire signage to you? What does it say?

EMERGENCY PLANS

All workplaces should have an emergency plan. The plan should detail what action needs to be taken by staff in the event of a fire, the evacuation procedure and what arrangements are in place for calling for the emergency services. In a small/simple workplace the plan may take the form of a fire-action notice positioned in conspicuous places where staff can acquaint themselves with the content. In larger/ more complex workplaces the plans will be more detailed and sophisticated.

It may be useful to aid the emergency plans with pictorial evacuation diagrams.

DISABLED/SPECIAL NEEDS PERSONS

If there are individuals working within your premises or likely to visit your premises that have special needs then your employer should have made specific or general provision for them, for example, audible alarms for the visually impaired or flashing lights for those with hearing difficulties. The action taken will be dependent on the workplace, the resources available to your employer and the specific needs of individuals.

PERSONAL EMERGENCY EVACUATION PLANS (PEEPs)

Where there are specific needs for individual employees then the employer must develop a personal emergency evacuation plan (PEEP) for that individual based on his/her needs. If you work with or near someone with a PEEP, you may need to acquaint yourself with his/her plan as one day you may need to assist him/her.

GENERIC EMERGENCY EVACUATION PLANS (GEEPs)

Where visitors may visit your premises, employers may need to develop a number of generic emergency evacuation plans (GEEPs) covering the possible scenarios in readiness for those situations. You may need to acquaint yourself with those plans as one day you may need to assist them.

SAFE REFUGES

These are places within your workplace designated to be suitable for persons to wait within the building to be rescued if it is not possible for them to evacuate the building safely. They must only be designated where the employer is satisfied that the location is safe and suitable to be designated as a safe refuge and that it meets the necessary fire safety criteria.

Safe refuge point

MEANS OF ESCAPE

All designated means of escape must be as short as possible and lead to a safe final exit point or other point of safety.

The escape route (including stairwells) should be wide enough for people to travel comfortably and they should be clear of obstructions and trip hazards.

Max travel distance*	More than one exit route	Single escape route
Low fire risk	60m	45m
Normal fire risk (factory)	45m	25m
Normal fire risk	45m	18m
Normal fire risk (sleeping)	32m	16m
High fire risk	25m	12m

* To an exit, for example open air, safe refuge

DOORS

Fire exit doors should ideally open in the direction of travel but this may not always be practical or feasible. Your employer should determine which doors really need to open out in the direction of travel based on their fire risk assessment. All fire doors should have self-closing devices and a good seal around them to prevent smoke travel. Fire doors should not be propped open unless they have an automatic self-closing mechanism which will release them in the event of an alarm activation. Fire doors are generally rated to provide protection for a predetermined amount of time. Their properties should not be compromised by the fitting of door furniture or vision panels (unless fitted correctly).

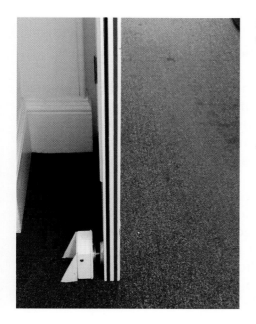

Never block automatic door closers

Are fire doors being propped open where you work? If so why?

FIRE DRILLS

Fire drills are generally held to:

- help employees become familiar with the action necessary in the event of a fire;
- help employees become familiar with escape routes;
- help employees become familiar with the alarms;
- highlight deficiencies in the evacuation process.

Employers should carry out as many drills as are necessary to achieve the above. Generally at least two evacuation practices should be carried out within the workplace annually but employers must take into account work patterns and risk when deciding how many are necessary. Generally it should be possible to evacuate most workplaces to a place of safety within 2–3 minutes. Employees must take all evacuation practices seriously and act as if they were in a real fire situation.

Fire bell

Q

When was the last time your employer carried out a fire drill?

ASSEMBLY POINTS

Assembly points should be established for all workplaces. The assembly points should be situated in a location which is unlikely to be affected by a fire and where assembled staff can safely leave the premises (if the assembly point is on site). Assembly points should not create entrapment spots and consideration should be given to prevailing wind directions when deciding on locations.

In the event of a necessary evacuation, employers should ensure that all staff can be accounted for, either by carrying out a roll call or an equivalent alternative means.

LIFTS

Never use a lift in the event of a fire unless it has been specifically designed to be used in a fire.

> **WARNING**
>
> IN THE CASE OF FIRE DO NOT USE THE LIFT. IF THE EMERGENCY ALARM IS SOUNDING, MAKE VOICE CONTACT WITH THE LIFT OCCUPANT(S) AND REPORT THE LOCATION AND NATURE OF THE EMERGENCY BY DIALLING **2222** FROM THE NEAREST TELEPHONE.

Do not use lifts in the event of a fire

EXERCISE 2

Self-assessment questions

To assess your level of understanding please complete the following exercise.

1. What are the two most common types of fire detector used?

   a) ....

   b) ....

2. Give two examples of a manual fire alarm system:

   a) ....

   b) ....

3. What type of fire extinguisher would you typically use on a small paper-based bin fire?

....

....

....

4. What is a class C fire?

....

....

....

5. Do all workplaces have to install emergency lighting?

   a) Yes
   b) No

6. What is a PEEP?

....

....

....

7. Is it generally wise to use a passenger lift within a building when the fire alarm sounds?

   a) Yes
   b) No

You will find the answers on page 80.

MODULE THREE

Fire law and risk assessment

A NURSING HOME IN LONDON IS FINED £200,000 WITH COSTS OF OVER £30,000 FOR CONTRAVENTIONS OF FIRE SAFETY REGULATIONS FOLLOWING A FIRE

In this module you will learn more about the Law relating to fire and fire-risk assessment.

The law relevant to fire is the Regulatory Reform (Fire Safety) Order (RRFSO). It applies to all non-domestic premises (except offshore installations, ships, land forming agricultural or forestry operations, means of transport, mines, bore hole sites and sports grounds).

The RRFSO is generally enforced by the local fire authority, although the Health and Safety Executive (HSE) and Environmental Health Officers (EHOs) can also enforce elements of fire safety legislation.

Failure to comply with the requirements of the RRFSO can lead to enforcement action being taken, which could include:

- **the requirement of Alteration notices;**
- **enforcement notices being issued;**
- **prohibition notices being issued;**
- **prosecution of the organisation/individuals.**

The order places duties on the responsible person, that is the person who is:

- the employer within a workplace to the extent they have control;
- any other person who has control over the premises;
- the owner of the premises.

The responsible person has to consider the following:

- reduction of the risk of fire and fire spread;
- the means of escape;
- keeping means of escape available for use;
- measures necessary for fire fighting;
- measures necessary for fire detection and warning;
- action to be taken in the event of a fire;
- training and instruction for employees.

In addition there are specific duties to:

- carry out suitable and sufficient risk assessment;
- review the risk assessment;
- record significant findings;
- manage fire risk.

FIRE-RISK ASSESSMENT – PURPOSE AND VALUE

A fire-risk assessment is an organised and methodical look at a premises and activities carried out to determine whether a fire could start and what existing controls are in place to minimise that risk and to help identify any additional controls necessary to prevent a fire starting or spreading. It is normally carried out by the employer (directly or through others); however, there is no reason why you should not carry out your own informal assessment wherever/whenever you are working. It could save your life or the lives of others.

A fire **Hazard** is something with the potential to cause harm through fire and/or explosion, for example substances, machines or methods of work.

A fire **Risk** is the probability of a fire or explosion occurring and the outcome if it does.

There are five steps to carrying out a fire-risk assessment:

Step 1: identification and control of fire hazards

Consider what is likely to cause a fire or explosion within your workplace or home. It is important to identify sources of heat/ignition, fuel and oxygen and consider how they may come together in an uncontrolled way resulting in a fire or explosion. Remember the hazard could include the way people are working.

Step 2: who could be harmed?

Consider who is likely to be at risk if a fire breaks out or if there is an explosion.

It could be fellow employees, visitors or customers in your workplace, your neighbours or even passersby.

You also need to consider:

- how they will be warned;
- how they will escape;
- where they work;
- permanent work stations;
- occasional work stations;
- contractors and temporary workers.

Step 3: evaluating the risks

Once the hazards have been identified the risks associated with them must be evaluated – often a score-based probability/consequences matrix is used to help prioritise risks. Where identified risk scores are not acceptable, additional controls need to be identified to reduce the scores to an acceptable level.

When evaluating risk consider the following:

- reducing sources of heat/ignition;
- reducing sources of fuel;
- reducing sources of oxygen;
- fire detection and fire warning;
- means of escape (MOE);
- means of fighting fire;
- maintenance and testing;
- fire procedures and training;
- disabled people and access;
- awareness training;
- other practical measures.

PRIORITY INDICATOR

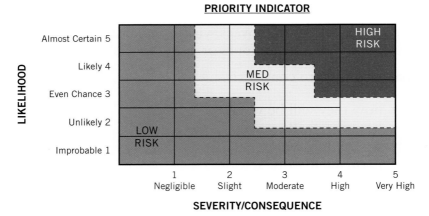

Priority indicator can help prioritise risk

Step 4: record keeping

Once the assessment has been completed, employers need to ensure that the significant findings have been recorded. The records can be paper based or in electronic format. It is important that employers also consider:

- emergency plans;
- information and instruction for employees;
- training of employees.

Step 5: review and revise

Once completed the fire-risk assessment needs to be reviewed from time to time.

Typically when there:

- are any changes in the workplace;
- are changes to a process;
- are changes to furniture, equipment, machinery, substances, buildings and so on;
- are changes to the number of persons employed or on the premises;
- are any new or increased hazard or risks;
- are changes in control measures;
- is a fire or "near miss" occurrence.

THE IMPACT OF FIRES ON THE ENVIRONMENT

Employers often forget to consider the impact large fires will have on the environment. Employers must always consider what impact any significant fire would have on the environment (in terms of land, air or water contamination) as well as the impact of the extinguishing media and its subsequent disposal.

THE IMPACT OF FIRES ON THE ORGANISATION

Employers often forget to consider the impact large fires will have on their organisation. Employers must always consider what impact any significant fire would have on the organisation (in terms of business interruption, staff morale, loss of stock, inability to service customers, reduced cash flow). Many businesses never recover from the impact of a significant fire.

EXERCISE 3

Self-assessment questions

To assess your level of understanding please complete the following exercise.

1. What does RRFSO stand for?

....

2. Who is responsible for ensuring a fire-risk assessment is carried out?

....

....

....

3. Give two examples of a responsible person:

a) ....

b) ....

---

4. Give three examples of situations that would require a review of the fire-risk assessment:

a) ....

b) ....

c) ....

5. Who normally enforces the RRFSO?

....

....

....

6. What enforcement actions can be taken against organisations failing to comply with the RRFSO?

a) ....

b) ....

c) ....

d) ....

7. List three examples of the business impact a serious fire could have on an organisation.

a) ....

b) ....

c) ....

You will find the answers on page 81.

Flammable liquids and liquefied petroleum gas

HIGH STREET BAKER FINED £50,000 FOR A NUMBER OF CONTRAVENTIONS INCLUDING BLOCKED CORRIDORS AND A FIRE EXIT THAT WAS SHUT TIGHT WITH FOUR PADLOCKS

In this module you will learn more about the dangers associated with flammable liquids and gases and the action necessary for fire prevention when using them.

DEFINITIONS

Flammable liquids are those liquids with a flash point between 32°C and 38°C.

Highly Flammable liquids are those with a flash point lower than 32°C.

Flash point is the lowest temperature at which the liquid can vaporise to form an ignitable mixture in air. At the flashpoint, the vapour may cease to burn when the source of ignition is removed.

Fuel	Flash point
Ethanol (70%)	16.6°C
Petrol	−43°C
Diesel	>62°C
Jet fuel	>60°C
Paraffin	38–72°C
Vegetable oil	327°C
Biodiesel	>130°C

Spontaneous combustion/auto ignition is a phenomenon in which a substance unexpectedly bursts into flame without apparent cause. Many substances undergo a slow oxidation that, like the rapid oxidation of burning, releases heat. If the heat so released cannot escape the substance, the temperature of the substance rises until ignition takes place. Spontaneous combustion often occurs in piles of oily rags, green hay, leaves or coal; it can constitute a serious fire hazard

Liquefied petroleum gases are gases in their natural state (usually butane and propane) stored and transported as liquids under pressure.

STORAGE OF FLAMMABLE LIQUIDS

When storing flammable or highly flammable liquids always consider the following:

- Store the liquids in suitable closed vessels or tanks.
- Only store minimal quantities inside work rooms/ buildings.
- Use containers that can be closed to prevent vapour release.
- Store the containers in fire-resistant cupboards.
- Larger quantities should be stored in fire-resisting well-ventilated external stores.
- Larger storage tanks should be bunded.
- Stores should be marked "Highly Flammable".
- Restrict all ignition sources in the vicinity.
- Appropriate fire-fighting equipment should be accessible.
- Be aware of drainage channels and contamination potential.

USE OF HIGHLY FLAMMABLE LIQUIDS

When using flammable or highly flammable liquids always consider the following:

- Only use in well-ventilated areas/cabinets.
- Ensure that there is no smoking and there are no ignition sources around.
- Replace lids on containers as soon as possible.
- Use safety containers, that is, metal, self-closing, flame arrestors.
- Prevent/control spillages.
- Treat empty containers as full unless purged or ventilated.
- Fire-fighting equipment should always be present.
- Safe means of escape should always exist.
- Employees must report defects and comply with all safety/fire rules in place.

STORAGE OF LIQUID PETROLEUM GASES (LPGs)

When storing liquefied petroleum gases always consider the following:

- Cylinders, tanks and pipelines should be in the open air or in a ventilated fire-resisting structure.
- A purpose-built underground reservoir/receiver vessel is recommended.
- Only minimum quantities should be stored within work rooms.
- Containers must be marked "Highly Flammable LPG".

USE OF LIQUID PETROLEUM GASES (LPGs)

When using liquefied petroleum gases always consider the following:

- Only use in well-ventilated areas.
- Avoid use in or near drains or basements.
- Avoid ignition sources and ensure there is no smoking.
- Ensure that there is appropriate signage etc.
- The use of flame arrestors may be needed.
- Ensure cylinders are kept secure and upright.
- Check for any leaks (never use a flame to do this).
- Isolate cylinders when not in use.
- Fire-fighting equipment should always be present.

EXPLOSIONS

Explosions are defined as the rapid flame propagation throughout an area containing flammable vapours, gases and dusts when these are within their flammable limits.

Different types of explosion include:

- unconfined vapour cloud explosions;
- confined vapour cloud explosions;
- boiling liquid expanding vapour explosions;
- dust explosions.

Explosions can occur when:

- the dust or vapours are combustible;
- the fuel is within its flammable limits;
- the fuel is capable of becoming airborne;
- an ignition source is available;
- the dust is of a particular size.

CONTROL OF EXPLOSIONS

Explosions can be prevented/controlled by:

- controlling sources of ignition;
- creating inert atmospheres with little or no oxygen present;
- preventing the formation of dust clouds via good housekeeping, vacuum systems, the use of enclosures;
- introducing engineering controls such as plant maintenance;
- introducing procedural controls such as monitoring and permits to work;
- introducing explosion-relief vents;
- using explosion containment and suppression systems;
- considering the location and building materials of plant in relation to other buildings.

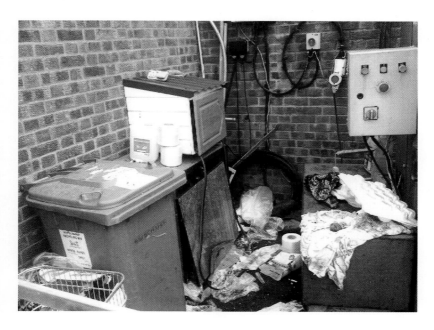

Store battery chargers away from rubbish

PERMIT TO WORK – HOT WORKS

In order to minimise/prevent the risk of fire/explosion when working it may be necessary to use a permit to work. A permit to work is a specific system designed to control high-risk activities (including hot works) and involves authorisations from specific individuals. Work should only be authorised where specific preconditions have been met and checks carried out.

Hot work can include:

- cutting;
- welding;
- brazing;
- soldering;
- the use of naked flame tools;
- drilling or grinding in flammable atmospheres.

Welding equipment

EXERCISE 4

Self-assessment questions

To assess your level of understanding please complete the following exercise.

1. What does HFL stand for? And list one example:

   a) ....

   b) ....

2. What does LPG stand for? And list one example:

   a) ....

   b) ....

3. What is the flash point of a liquid?

   a) ....

4. List three things that have to be considered when using HFLs:

   a) ....

   b) ....

   c) ....

5. List three things that have to be considered when using LPGs:

   a) ....

   b) ....

   c) ....

6. Is petrol more flammable than diesel?

   a) Yes
   b) No

7. List three examples of hot work:

   a) ....

   b) ....

   c) ....

You will find the answers on page 82.

Answers to questions

MODULE ONE – Fire

1. What three elements are needed to start a fire?

Heat/ignition, oxygen and fuel

2. By removing or controlling one of the elements what happens to the fire?

The fire will be extinguished

3. List three common causes of fire.

Arson, electrical equipment, electrical wiring, the use of naked-flame equipment, static electricity, gases, hot substances, explosions, discarded smoking materials

4. List three examples of sources of heat/ignition.

Grinding wheels, welding, naked flames, smoking materials, cooking appliances, heating appliances, electrical equipment/switches, hot surfaces, lighting, cooking, poorly ventilated equipment, static electricity or other similar answers

5. What are two of the dangers associated with fires?

Burns, smoke inhalation, toxic fumes, structural damage, environmental damage

6. List three examples of sources of fuel.

Wood, paper, cardboard, packaging, plastics, rubber, foam, textiles, waste materials, varnishes, paints, adhesives, petrol, white spirit, paraffin, methylated spirits, toluene, liquefied petroleum gases, acetylene, hydrogen or other similar answers

7. Give two examples of how fires can spread.

Conduction, convection, radiation, burning materials

MODULE 2 – Fire detection and protection

1. What are the two most common types of fire detector used?

Smoke, heat

2. Give two examples of a manual fire alarm system.

Word of mouth, the use of hand bells, whistles, hand strikers, rotary gongs, hand-operated sirens

3. What type of fire extinguisher would you typically use on a small paper-based bin fire?

Water

4. What is a class C fire?

One involving gases or liquefied gases

5. Do all workplaces have to install emergency lighting?

b) No

6. What is a PEEP?

Personal emergency evacuation plan

7. Is it generally wise to use a passenger lift within a building when the fire alarm sounds?

b) No

MODULE THREE – Fire law and risk assessment

1. What does RRFSO stand for?

Regularity reform fire safety order

2. Who is responsible for ensuring a fire risk assessment is carried out?

The responsible person

3. Give two examples of a responsible person.

The employer within a workplace to the extent they have control, any other person who has control over the premises, the owner of the premises

4. Give three examples of situations that would require a review of the fire-risk assessment.

Any changes in the workplace; changes to a process; changes to furniture, equipment, machinery, substances, buildings; changes to the number of persons employed or on the premises; any new or increased hazard or risks; changes in control measures; following a fire or "near miss" occurrence

5. Who normally enforces the RRFSO?

The local fire authority

6. What enforcement action can be taken against organisations failing to comply with the RRFSO?

Alteration notices, enforcement notices, prohibition notices, prosecution

7. List three examples of the business impact a serious fire could have on an organisation.

Business interruption, staff morale, loss of stock, inability to service customers, reduced cash flow, prosecution, reputational damage

MODULE FOUR – Flammable liquids and liquefied petroleum gas

1. What does HFL stand for? And list one example.

Highly flammable liquid; petrol, ethanol

2. What does LPG stand for? And list one example.

Liquid petroleum gas; propane, butane

3. What is the flash point of a liquid?

It is the lowest temperature at which the liquid can vaporise to form an ignitable mixture in air

4. List three things that have to be considered when using HFLs.

Use in well-ventilated areas/cabinets; ensure that there is no smoking and there are ignition sources around; replace lids on containers as soon as possible; use safety containers, that is, metal, self-closing, flame arrestors; prevent/control spillages; treat empty containers as full unless purged or ventilated; fire-fighting equipment; safe means of escape; employees must report defects and comply with all safety/fire rules in place

5. List three things that have to be considered when using LPGs.

Use in well-ventilated areas; avoid use in or near drains or basements; avoid ignition sources; no smoking; signs etc; flame arrestors; the security of cylinders; leak testing; isolate cylinders when not in use; fire-fighting equipment

6. Is petrol more flammable than diesel?

a) Yes

7. List three examples of hot work.

Cutting, welding, brazing, soldering, use of naked-flame tools, drilling or grinding in flammable atmospheres

Glossary/definitions

Bunding Containment around a vessel or tank to capture any spillage

Conduction Heat transfer across a metal object

Convection Heat transfer through hot air rising

EHO Environmental Health Officer

Fuel Something that will burn

GEEP Generic emergency evacuation plan

Hazard Something that has the potential to cause harm

HFL Highly flammable liquid

H&S Health and Safety

HSE Health and Safety Executive

LPG Liquefied petroleum gas

PEEP Personal emergency evacuation plan

Radiation Heat transfer through space

Risk The probability and consequences of a hazard occurring

SHE Safety, Health and Environment

Toxic Poisonous

UK United Kingdom

Examination

Once you have completed the guide you can choose to complete the examination.

Complete the following exam to test your fire safety knowledge. All candidates will be informed whether they have passed or failed via email, and candidates achieving a pass mark of 75% or above will receive a certificate.

To find out details of where to send your completed exam, or to request an electronic version, please visit www.routledge.com/9780415835428.

You have forty minutes – good luck

Your FULL Name: (PRINT)		Employee Reference:	Employer Name:
Date of Birth:		Your Email Address:	
Date of Examination:			

Please indicate your answer(s) by putting a tick in the appropriate box(es)

1. What three elements are needed to start a fire?

   a) Heat, fuel and oxygen ❒
   b) Heat, fuel and smoke ❒
   c) Fuel, heat and matches ❒
   d) Oxygen, fuel and nitrogen ❒

2. How do fires spread?

   a) Conduction and radiation ❒
   b) Convection and radiation ❒
   c) Burning material and radiation ❒
   d) All of the above ❒

3. What is a class C fire?

   a) a fire involving wood ❏
   b) a fire involving metals ❏
   c) a fire involving oils and fats ❏
   d) a fire involving gases ❏

4. What is meant by the term **PEEP**?

   a) Personal emergency evacuation plans ❏
   b) Personal evacuation exit plans ❏
   c) Plant emergency evacuation procedure ❏
   d) Personal emergency evacuation procedure ❏

5. What is the flash point of petrol?

   a) −43°C ❏
   b) +33°C ❏
   c) −73°C ❏
   d) −20°C ❏

6. Which one of the following **is not** a duty of a fire warden?

   a) Checking fire doors are kept closed ❏
   b) Carrying out roll calls ❏
   c) Entering burning buildings to find colleagues ❏
   d) Being familiar with the fire procedures ❏

7. Who generally **enforces** Fire Safety legislation in the UK?

   a) The fire and rescue authority ❏
   b) Policemen and firemen ❏
   c) Factory Inspectors and Environmental Health Officers ❏
   d) Environmental Health Officers and firemen ❏

8. Which of the following attributes is desirable in a fire door?

   a) Self-closing device ❏
   b) Large perspex vision panels ❏
   c) Large gaps around the door frame to allow airflow ❏
   d) Inward openings ❏

9. Is it a legal requirement to install emergency lighting in all work places?

   a) Yes ❏
   b) No ❏

10. What are the two most common types of fire detector?

    a) Smoke and light detectors ❏
    b) Smoke and heat detectors ❏
    c) Light and smoke detectors ❏
    d) Radiation and smoke detectors ❏

11. How can fires impact the environment?

    a) They provide oxygen for the environment ❏
    b) They light up the environment ❏
    c) They can pollute land, air and water ❏
    d) They improve the air quality ❏

12. What are the **main purposes** of fire evacuation drills?

    a) To identify failings in the evacuation procedures ❏
    b) To identify how quickly the building can be evacuated ❏
    c) To satisfy legal requirements ❏
    d) All of the above ❏

13. Which **factors** must be taken into account when carrying out a fire-risk assessment?

    a) Sources of heat, fuel and oxygen ☐
    b) The number of people on the premises ☐
    c) The presence of explosive substances on the premises ☐
    d) All of the above ☐

14. Which of the following **is not** a **danger** normally associated with fires?

    a) Death ☐
    b) Burns ☐
    c) Structural damage ☐
    d) Whole body vibration ☐

15. What does RRFSO stand for?

    a) Regulatory reform first standard order ☐
    b) Regular radio frequency state order ☐
    c) Regulatory reform fire safety order ☐
    d) Regulatory reform fire safety organisation ☐

16. Where might a hot work permit be used?

    a) Welding operations ☐
    b) Work at height operations ☐
    c) Naked flame burner operations ☐
    d) Confined space operations ☐

17. On which type of fire would you **not** use a water-based fire extinguisher?

   a) Paper-based fire ❑
   b) Wooden-furniture-based fire ❑
   c) Electrical-appliance-based fire ❑
   d) Textiles-based fire ❑

18. Could a bell or whistle be used to alert people of a fire?

   a) Yes ❑
   b) No ❑

19. What is the **definition** of a fire **risk**?

   a) Something that will cause harm ❑
   b) Something that has the potential to cause harm ❑
   c) The probability of a fire or explosion occurring and the likely consequences ❑
   d) Being struck by a burning door ❑

20. What is the **definition** of a fire **hazard**?

   a) Something that will cause harm through fire or explosion ❑
   b) Something that has the potential to cause harm through fire or explosion ❑
   c) The probability of harm being caused and the likely consequences ❑
   d) Being struck by a burning door ❑

Well done, you have now finished the test. Please check to ensure that you have answered all questions (remember some questions require more than one answer).